HF731894

Paul Doherty is one of Britain's foremost artist-astronomers. Assistant Director of the Saturn Section and Co-ordinator of the Mercury Section of the British Astronomical Association, he works as a freelance illustrator and writer on astronomy, space, geology and the sciences. In the last two years he has illustrated 23 books on these subjects, including *Stargazing: Astronomy Without a Telescope* by Patrick Moore (Barron's, 1985). He is a regular illustrator for the BBC's *Sky at Night* programmes. He has written many magazine articles, and his book *Atlas of the Planets* was published in 1980.

Patrick Moore is known to an audience of millions through his popular television series, *The Sky at Night*. He has written over sixty books including *Stargazing* (1985), *Guide to the Stars* (1977), *Guide to the Planets* (1976), *Atlas of the Universe* (1970), *The Amateur Astronomer* (1970) and *Moon Flight Atlas* (1969). He is President of the Astronomical Association and Director of the Armagh Planetarium. He has received numerous gold medals for his services to astronomy, and in 1968 was made an Officer of the Order of the British Empire.

Ray Bradbury is one of the world's leading science fiction writers. His celebrated novels, short stories and plays have helped to bring the science fiction genre to the forefront of serious contemporary literature. His novels include *The Ghosts of Forever* (1981), *Mars and the Mind of Man* (1973) and *I Sing the Body Electric!* (1969). Much of his work has been adapted successfully for television and the cinema, including *Fahrenheit 451* (1953), *The Illustrated Man* (1951) and *The Martian Chronicles* (1950).

THE ARRIVAL OF HALLEY'S COMET 1985/86

PAUL B. DOHERTY

Foreword by RAY BRADBURY
Introduction by PATRICK MOORE

BARRON'S
Woodbury, New York/London/Toronto/Sydney

Photograph on previous page: Halley's Comet at its last return. This photograph was taken by Knox-Shaw on 10 May 1910 with the 750mm (30 inch) reflector at Halwan Observatory, Egypt. (Courtesy Royal Astronomical Society)

First Edition published in the United States, its dependencies, the Philippines, and Canada 1985 by Barron's Educational Series, Inc.

Copyright © Paul B. Doherty 1985

All rights reserved. No part of this book may be reproduced in any form, by photostat, microfilm, xerography, or any other means, or incorporated into any information retrieval system, electronic or mechanical, without the written permission of the copyright owner.

All inquiries should be addressed to:
Barron's Educational Series, Inc.
113 Crossways Park Drive
Woodbury, NY 11797

Library of Congress Catalog Card No. 85-15089

International Standard Book No. 0-8120-3632-8

This book has been designed and produced by Aurum Press Limited, 33 Museum Street, London WC1A 1LD, U.K.

Printed in Great Britain
56789 987654321

CONTENTS

ACKNOWLEDGEMENTS

My thanks are due to Patrick Moore for writing the introduction to this book and to Ray Bradbury for his foreword. To Michael Haggiag of Aurum Press for his support. To my wife Linda and to Mrs Gwen Gater for their help with the manuscript.

HALLEY & TWAIN: THEIR COMET
by Ray Bradbury

A few years ago, crossing the Atlantic by ship, I looked up at the night sky and wrote a poem the thrust of which was:

> They have not seen the stars!
> Not one, not one of all the beasts
> Who graze in meadows, sit on hills
> Has looked, looked up
> And seen what we have seen.
> The stars, oh Lord, much thanks, the stars!

Since then I have been given cause to think that most humans, myself included, have forgotten to do just that, quietly celebrate each night that is, in reality, a great fireworks, the Universe's eternal July 4th. We get so busy with non-living that we forget to breathe, to note our personal temperature, to touch, to hear, but above all to see.

Halley's Comet comes back to visit and remind us of just that, plus a panoramic wide-screen reminder of our obligations toward Space, which we have been shirking more often than not in the past few years.

It also reminds us of the sad fact that the United States has forgotten to dream tall and fly even taller in connection with this very Phantom of Time, or Ghost of the Abyss, or however else you might want to name Halley's pale son.

We in America had a chance to toss our souls up, via solar sail, to meet and for an instant pace the Comet. We decided against it. In that moment, Mark Twain must have turned with a groan in his tomb. For it was old Mark, you surely recall, who commented on the fact that he was born in the year of the Comet back in the nineteenth century, and that he would die with the next visitation of the spectral prowler in 1910. Whereupon he did just that.

I suggested, a few years back, that we name our mechanical encounter with the Comet after Mark Twain. What a fine steamboat name to churn up for greetings and farewells, representing us and all the miracle-beholding sons and daughters of Earth. But it was not to be.

So, others will probe for us, speak instead of us, see better than us. So perhaps any probe from Europe should be named Jules Verne, or at the very least, Passepartout or Phileas Fogg, in honor of the huge traveller's journey Around the Universe in (almost) 80 Years.

No matter, now. The deed is done. We have chosen not to travel first class, but to sling ourselves under some foreign caboose on the way to Visitations.

What Halley would have thought of our anxieties and joys about his overgrown child, I cannot say.

But Mark Twain had an opinion on everything and I dare to speak for him. Next year, Mr Twain will be touched awake by the stellar ghost and will say something like this:

> Is it time to get up? All that light put me in mind of dawn. But I see it is only my old portent that waked me as a babe and slept me as doddering fool. If the dogs and cows and the birds in meadows and trees cannot see it, and if the damned time-bandits in cities, who waste the time they have pilfered, refuse to look up, well, then, let us, you and I, sit on the sill of my tomb and stare until we are blind with tears that welcome a great and beautiful dream. If there are some few boys born on the day when Halley's vision is first perceived in this year 1986, let one be named for me, do you mind? And let him live to the length of the Comet's circuiting and pass, as I passed, when it returns. Now let us be quiet. And let us stare.

And so say I. And shut this preface. But to open my eyes.

INTRODUCTION
by
Patrick Moore

Halley's Comet is one of the most famous of all cosmic visitors. This is all the more surprising because at the time when I write these words (May 1985) there are relatively few people living who have actually seen it. The last return was that of 1910, and before that the comet had not been on view since 1835.

Halley's Comet is an old friend, inasmuch as records of it go back for thousands of years. It has not always been regarded as benign; there was a time when comets caused a great deal of alarm and despondency — even in 1910 there was some evidence of this. We know better today. It is a matter for regret that our opportunities are so limited. The comet comes back to the Sun only once every 76 years, and that is a long time.

For most people, then, there is only one chance in a lifetime to see Halley's Comet. Of course it is not a question of 'one night only'; the comet should be a naked-eye object for some time in late 1985 and early 1986, but this is admittedly only a brief part of the comet's revolution period of 76 years. Let us make the most of it.

Comets, of course, are common enough. Many are known to revolve round the Sun, and several are always under observation at one time or another. What, then, makes Halley's Comet so special? Simply that it is the only bright comet which we can predict. All the other really spectacular comets — such as those of 1811, 1843, 1858, 1882 and so on — have periods of so many centuries that as far as we are concerned they might just as well be non-periodical. This also applies to other comets which have become famous — or notorious! Such was Kohoutek's Comet of 1973, which promised to become really brilliant, but which signally failed to come up to expectations. Whether it will do better at its next return is a matter for debate, but it is of academic interest only, since it will not be on view for about 75,000 years.

I think that the case of Kohoutek's Comet shows that our knowledge of comets as a whole is still very incomplete. But at least Halley's Comet is more reliable and we have already a very good idea of what it will look like, even though it may have major surprises in store for us during the next few months.

Last time Halley's Comet was close to us, in 1910, the return was much more favourable than the present one (even though it had been to some extent upstaged by the brighter Daylight Comet, which had shone down in January 1910, two months before Halley; in fact, most people who believe they saw Halley's Comet in that year really saw the Daylight Comet instead). Yet the current return is much more important, because we have the chance of carrying out studies with equipment of the kind which was quite undreamed-of in 1910. It is not only a question of telescopes and spectroscopes; we now have all manner of sophisticated instruments, and, of course, we can send spacecraft to rendezvous with the comet.

This is a chance not to be missed, because only Halley's Comet gives us the opportunity we need. Though it remained unseen between 1911 and its recovery in 1982, we have always known its exact position, and we have been able to make plans well in advance. The Americans have missed their opportunity — their Halley probe was cancelled on the grounds of cost, which seems an almost incredible decision (if you doubt me, compare the cost of a comet probe with that of half a nuclear submarine). However, five probes remain: two Russian, two Japanese and one European. The Russian and Japanese probes will bypass the comet at fairly close range. The European spacecraft, Giotto, will — we hope — actually penetrate the comet's head, and send back direct pictures of the nucleus, about which we still know very little.

This is, in fact, a very nice piece of international co-operation. The International Astronomical Union, the controlling body of world astronomy, is one of the very few organizations which is genuinely international; there are no Iron Curtains, Bamboo Curtains or curtains of any kind. With Halley's

Comet all the nations involved are working together. Let us hope that this sets a trend for the future.

There are many things the observer has to contend with these days, all of which seem designed to make an astronomer's life more difficult. There are ways to combat them, though. On many occasions older people have complained to me that the stars these days don't seem so bright compared with those of their youth. Obviously the stars have not changed; they remain the same for thousands, even millions of years. Certainly no change is likely in one's lifetime. The remark is not without foundation, however. The conditions in which we see the stars have certainly changed. Most damaging are the many artificial lights we now have. People who live in cities are lucky if they see even the brightest stars. Only those who travel to the countryside can appreciate the full beauty of the night sky. Here the stars stand out like jewels. This sort of thing will affect the visibility of even the brightest comets: only when miles away from city lights will their majesty be seen, only then will the delicate streamers of dust and gas which make the spectacular tail be revealed.

Comets may be minor members of the solar system, but they are very important, and we rely upon the coming return of Halley's Comet to increase our knowledge tremendously. It is, therefore, hardly surprising that many books on the subject are being produced. Paul Doherty's fame as an artist is so great nowadays that one tends to forget that he is a highly competent astronomer in his own right, and an official member of the International Halley Watch. In this short book he has admirably summarized the main facts about the comet: its history, its nature, what we may expect to find out within the next year, and how to set about observing it, either visually or with a camera.

I recommend the book to you without hesitation. Read it, learn from it — and then enjoy the few months when Halley's Comet will shine down from us before departing again for the far reaches of the Solar System.

WHAT ARE COMETS?

The appearance of a brilliant comet in the sky is one of the most magnificent sights astronomy has to offer. Comets have always impressed us, be it with fear or with awe.

In days gone by they were regarded as portenders of disaster. Tales of famine, pestilence, wars and the deaths of important people were associated with the appearance of comets. The very word disaster (*dis* — evil, *aster* — star) speaks for itself.

This attitude persists even today, despite our increasing knowledge of comet science. Three recent cases illustrate this. In 1970 an interesting and beautiful comet was seen during March and April, a conspicuous naked-eye object. At the time the Arab-Israeli conflict was heated, and many Arabs thought the comet to be some sort of secret weapon designed by Israeli scientists. More recently the appearance of the much heralded Comet Kohoutek prompted some religious sects to suggest that the comet signified the end of the world. Finally, in 1983 a small comet made a very close approach to Earth and despite its size was easily visible to the naked eye as a large round misty patch. Shortly before this a book written by two well-known researchers, Sir Fred Hoyle and Professor Chandra Wickramasinghe, had been published. In the book it was suggested that diseases originate in outer space and that material left behind by comets could add to this problem. This was picked up by the popular press at the time the comet made its approach, and proclaimed with headlines such as 'E T PLAGUE RISK FROM NEW COMET' and 'COMET COULD BRING KILLER SPACE BUGS'. The Theory of Hoyle and Wickramasinghe has been heavily criticized, but it contains many valid points and certainly does not deserve to be treated in such a sensational fashion.

There have been many brilliant comets down the centuries, and records tell of some that were of spectacular proportions, with tails that stretched right across the sky and heads brighter than the full Moon. It is tempting for us to be sceptical of these reports. Of late there has been a dearth of bright comets, compared with earlier times. In the fifteenth and sixteenth centuries there was a swarm of them. Brilliant naked-eye comets appeared in 1402, 1403, 1449, 1456 (Halley), 1457 (two), 1472, 1500, 1506, 1531 (Halley), 1532 (two), 1533, 1538, 1539, 1556, 1558, 1577, 1580 and 1582. In the nineteenth century at least eight brilliant comets were reported. In 1811 there appeared a comet of amazing size: its head was about the size of the Sun and its tail was estimated to be in the order of 100 million miles long (160 million km). In 1843 another comet appeared which had the longest tail ever recorded, and was approximately double the length of the 1811 comet — so long in fact that if the head of the comet had been placed at the position of the Sun the tail would have reached out further than the orbit of Mars. The head of this comet was easily visible in daylight.

In 1858 Donati's Comet, though not particularly great, was far more spectacular than anything we have had this century. It was regarded as the most beautiful comet ever seen. In 1861 Tebutt's Comet passed so close to Earth that for a while we were actually inside it.

In the latter part of the nineteenth century bright comets came thick and fast. There was a brilliant one in 1862, Coggia's Comet in 1874, two great comets in 1880 and 1882 (the latter of which was the first to be photographed by Sir David Gill) and finally one in 1887.

So far during the twentieth century we have seen only a few bright comets, and none could really be termed great. Most impressive was the Daylight Comet of 1910, and of course Halley of the same year.

The Arend-Roland Comet of 1957 put on a good show, as did Ikeya-Seki in 1966. Bennett's Comet of 1970 was a conspicuous sight during the early hours of the morning through the last week of March and the first two weeks of April. Best of all was West's Comet of 1976. At its brightest in the first few days of March it could still be seen in bright twilight. It had a beautiful fanned tail which was about 12 degrees long, or 24 times the apparent diameter of the full Moon.

Though it became famous, Comet Kohoutek was a bitter disappointment to those who thought it would be the comet of the century. The press went overboard in their coverage of the comet. It had been discovered at a great distance from the Sun and there was every reason to believe that it would be spectacular, but although very interesting scientifically it was not an easy naked-eye object. Following its failure to live up to expectations no coverage was given to Comet West, which was unexpectedly impressive. Unfortunately it slipped by virtually unnoticed by the general public.

It is little wonder therefore that Halley is awaited with such excitement.

What then are these unusual objects that cause so much interest? Their proportions in some cases are truly gigantic, yet enormous though they are, the material they contain is so small as to be all but non-existent. Comets could be termed the spectres of the solar system.

The actual solid body of the comet, the nucleus, has never been satisfactorily observed and our ideas about it are still largely speculative. In even the largest and brightest comets it is never seen as anything but a tiny star-like point. Estimates of the size of the largest nucleus do not exceed 6 km (4 miles), with 1 km probably being more likely. How can something so small give rise to so spectacular an effect?

It must be the material which goes to make up the nucleus that is responsible. Present theories suggest that the nucleus is a small piece of rock mixed with ices — or, as one astronomer, F. Whipple, described it, a 'dirty snowball'. Only when this object is close to the Sun does activity begin and the phenomena that go to make the bright object we see are triggered off. The ices in the nucleus start to evaporate, forming a cloud of gas. It is this we call the

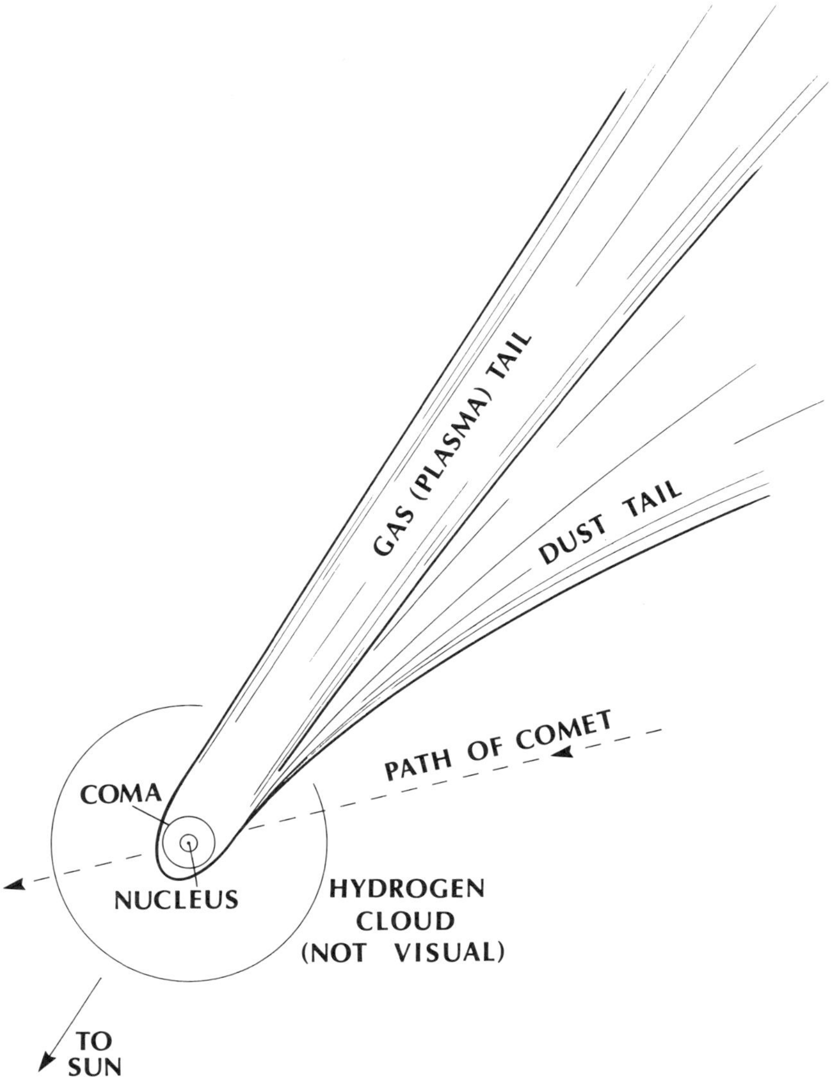

THE STRUCTURE OF A COMET
The main features of a comet, including the Type I (gas) and Type II (dust) tails. The large hydrogen cloud surrounding the coma is not a visual structure. Its existence was first predicted in 1968 and its eventual observation was made by the Orbiting Astronomical Observatory (OAO) 2 while studying the comet Tago-Sato-Kosaka 1969g on 14 January 1970.

Further observations of Comet Bennett 1969i by OAO2 and OAO5 on 5 February 1970 confirmed the existence of this huge cloud.

The diagram does not show this feature to scale. It can have a radius of between 1 and 10 million km (600,000 and 6 million miles).

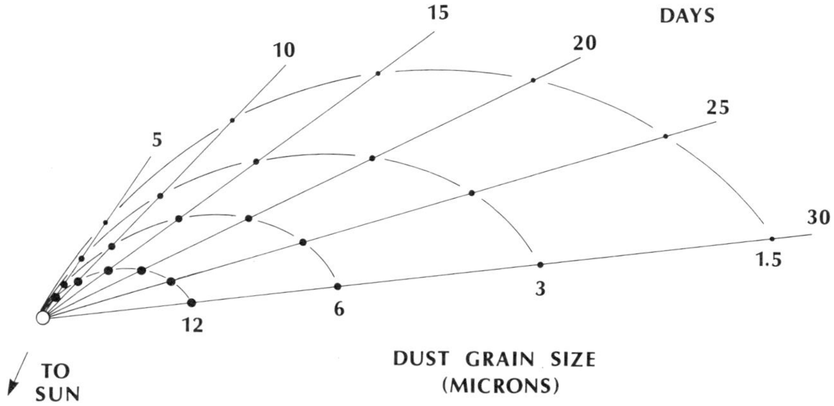

WHY A DUST TAIL CURVES
Comet tails always point away from the Sun, and some are markedly curved. This curving is most often seen in the Type II dust tail, because the comparatively large particles making up this type of tail are less drastically affected by solar wind than the tiny particles of the gas tail and leave the comet head at lower speeds. The larger the particles the more slowly they leave the head and the more they appear to lag behind. This curving is most noticeable around perihelion when the speed of the comet is greatest and its direction of travel is at right angles to the direction of the solar wind pressure.

(a)

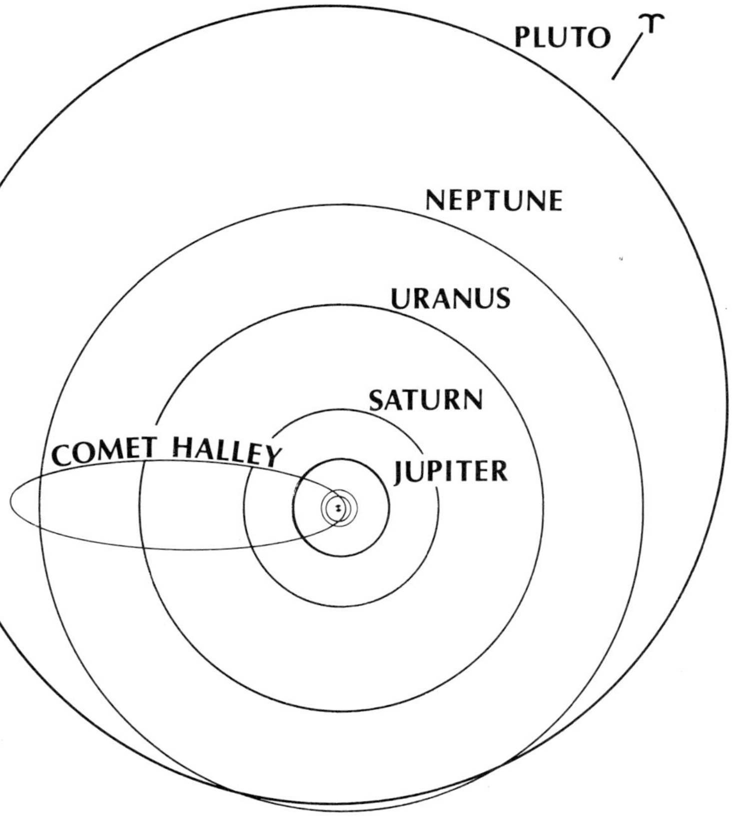

THE ORBIT OF HALLEY'S COMET
Halley's Comet orbits in a long drawn-out cigar-shaped ellipse, making the orbits of the planets, which are only slightly elliptical, appear almost circular (a).

It can be seen that the orbit of the comet crosses the orbits of several of the planets, and one might wonder why a collision never occurs. A glance at (b) will show why. Halley's orbit is highly inclined to the ecliptic (the path of the Earth around the Sun), and since all the planets, with the exception of Pluto, are less highly inclined no collision is possible — at least so long as Halley's orbit remains as it is today.

Diagram (c) shows the dates when Halley's Comet crossed the orbits of the planets during its last orbit.

(b)

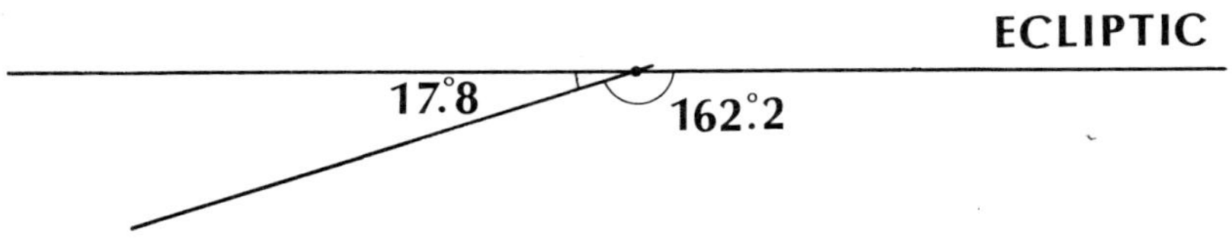

(c)

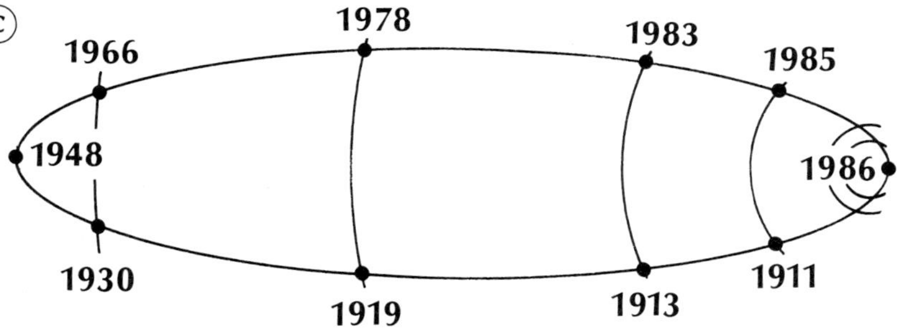

'coma'. As the comet gets nearer to the Sun the evaporation rate increases. The coma becomes larger and brighter.

Eventually these gases are affected by solar radiation pressure, or solar wind. This is a moving stream of electrically charged particles from the Sun, and it is this that deflects the gases and particles in the coma of the comet. A tail is formed by the gases being driven from the head of the comet in a direction away from the Sun.

Detailed views of the heads of brighter comets reveal many jets of gas and dust emanating from the nucleus. It is thought that the nucleus may have a thin crust, possibly of silicate. The gases probably escape through fissures in this crust, or through holes caused by impact with other smaller bodies.

With increased activity around the nucleus, dust as well as gas is given off. The more dust, the more spectacular the comet — and some comets are very dusty indeed.

At this stage some comets exhibit two types of tail. What is known as the Type I tail is a straight filamentary plasma or gas tail which tends to be bluish in colour and shines because of ionization. This type of tail is affected most strongly by the solar wind; material leaves the nucleus at great speed and so the tail is always quite straight.

There is also a fairly featureless dust tail or Type II tail, which sometimes becomes gently curved. This is due to the comet's rapid motion around the Sun and the relatively slow speeds of the particles of dust as they leave the comet. Large dust particles leave the comet more slowly than the smaller particles, thus causing a curved effect (see diagram on p.12).

Scientists now have a reasonable understanding of the material that makes up a comet. Observations using a spectroscope tell us about the various gases in the comet. A continuous spectrum is obtained from the dust tail, which shows that this is the result of sunlight reflected by the solid particles within it.

The gas tail gives an emission spectrum or bright line spectrum, informing us that it is gas ionized by solar radiation. This gas is excited into luminosity in much the same way as the gas in a fluorescent tube becomes luminous when an electric charge is passed through it. Inside the fluorescent tube is a gas in a near-vacuum. When charged it shines brilliantly, which is why comets look so dramatic even though they are so rarefied.

By examining the lines in the spectrum we can find out what gases are present. Gases shine in various wavelengths, and by comparing the spectrum of a comet with spectra of known gases we can identify those present in the comet. Most common of the gases found are cyanogen (C_2N_2), carbon (C_2), methane (CH_4), ammonia (NH_3), carbon dioxide (CO_2), carbon monoxide (CO), nitrogen (N_2) and sodium (Na). Compounds of oxygen and hydrogen have also been found.

Why do the comets come and go so quickly and where do they come from? Comets orbit the Sun just the same as the planets, but their orbits, instead of being nearly circular like the orbits of the planets, are long and drawn out ellipses which swing in very close to the Sun. They spend long periods of time at great distances away, travelling very slowly. It is only for a short part of their orbit, when they are close to the Sun, that we see them.

There are probably many millions of comets. A popular theory put forward by J. Oort of Leiden University suggests that surrounding the solar system there is a vast cloud of comets way out beyond the most distant planets, 40,000 times as far from the Sun as is the Earth. The Oort Cloud, as it is now known, may contain as many as 100 billion comets. Despite their distance they are still gravitationally part of the solar system, and it is thought that every now and then the passage of stars close to the cloud might disturb some of these comets, causing them to fall towards the inner regions of the solar system.

If during this journey a comet should pass close to one of the major planets, the gravitational pull of that planet might modify the comet's motion, throwing it into a short-period orbit. If no such approach to a planet occurs the comet will return to the cloud, probably not revisiting the inner solar system for thousands, even millions of years.

There are two basic types of comet, the periodic and the non-periodic. Periodic comets return to the Sun regularly and have been observed at more than one return. Those termed non-periodic probably do have a period, but this is so long that only one return is on record.

Halley returns regularly, with a period of around 76 years. There are a number of comets with short periods in the order of 4–7 years, though none are as bright as Halley. It seems likely that shorter period comets have been in some way captured by the planets and gravitationally locked into smaller orbits. The gravitational effects of the planets also change the orbital period of Halley, either speeding it up or slowing it down. For this reason Halley's period ranges from 75 to 80 years.

COMET HALLEY'S HISTORY

There are many reasons why Halley's Comet is so important. The most obvious is that it is the brightest of the shorter period comets and was the first to be recognized as periodic. Another reason is that since comets are believed to be left over from the time when the solar system was formed — about 4500 million years ago — by studying them we hope to learn more about how the solar system itself came into being. In the case of Halley we know of its coming well in advance and so can prepare for it, unlike many bright comets which take us by surprise.

In 1680 and 1682 the English astronomer and mathematician Edmond Halley (later to become Astronomer Royal) observed two bright comets which particularly impressed him. At first he tried to calculate the orbit of the 1680 comet, finding that its path resembled those of other recorded comets. He had the idea that this comet returned to the Sun after periods of 575 years. This actually turned out to be incorrect, since the comet has a period in the order of many thousands of years. But the idea was a good one, and when he applied it to the comet of 1682 things started to fall into place. He found the orbit of the comet remarkably similar to other comets of 1607 and 1531. Further historical checks showed the comets of 1301, 1378 and 1456 to have been very similar. All had the same inclination to the ecliptic of around 17 degrees (see diagram on p.13), all had the same perihelion distance (closest approach to the Sun), and all had a retrograde motion — that is, they all travelled around the Sun in the opposite direction to the planets. Could they be the same comet? In 1705 he made the prediction that the comet's period was around 76 years and that it would therefore return in 1758. Halley of course did not live to see its return, but when he died in 1742 he was confident that his prediction was correct.

The comet was indeed picked up on Christmas night

1758 by a German amateur astronomer, Johann Palitzsch. Halley was right, and suitably it is his name that the comet carries. It is usual for a comet to bear the name of the discoverer, but there are occasions when the person who computes the orbit has the comet named after him. In the case of Halley's Comet we have no idea who first saw it, so it is only fitting that it should be named after the person first to realize that it was periodical — and first in fact to prove that comets did return to the Sun.

It is possible to compute the dates of previous returns, and this has been done as far back as 2646 BC. It has only been possible thus far to correlate actual observations with the computed dates as far back as 240 BC, with the possibility of two other sightings in 466 BC and 1059 BC. There have been efforts to correlate more, but the further back one goes the more uncertain the records become.

From the following list it may be seen that only records from the last 1900 years are really reliable. Things become very uncertain before about AD 141. Added to the list are events of particular interest: history has been very colourful at the various returns of Halley, which is yet another reason why the comet holds such interest for the layman as well as the astronomer.

We can be fairly sure that the comet has been around for many thousands, even millions of years. One conservative estimate suggests that it has been paying regular visits for at least 200,000 years and will probably continue to do so for at least a comparable amount of time. Many scientists believe that comets are left over from the very creation of the solar system, from which we can expect Halley to have been around for a few thousand million years. Its orbital period is likely to have been considerably modified in the more recent past, but all of this is speculation.

RECORDED RETURNS OF HALLEY'S COMET

1910 20 APRIL

Recovery announced by Max Wolf at Heidelberg, Germany. Image found on photographic plate taken as early as 24 August 1909. The distance of the comet from Earth was then 770 million km (480 million miles). Observed over a period of 22 months. Last photographed on 30 May 1911 from Lowell Observatory, Arizona. This was the first return when photography could be applied.

The Earth was thought to have passed through the comet's tail on 18 May 1910 and a close approach to the comet (21 million km, 14 million miles) occurred on 20 May. A very favourable return.

Even here the association of comets with ill omens prevailed. Some people connected this return with the death of King Edward VII of England which occurred in the same year.

1835 16 NOVEMBER

Recovered by Father Dumouchel and Francisco di Vico at Collegio Romano Observatory on 6 August 1835. Last seen by John Herschel on 20 May 1836. Visible in all for 9½ months. Close approach to Earth on 10 October (7.5 million km, 4.6 million miles). Very favourable return.

1759 13 MARCH

Recovered by Johann Palitzsch on 25 December 1758. Also observed by Charles Messier. This was the first predicted return. Well seen though never brilliant. Last seen 22 June 1759 after 6 months' observation.

1682 15 SEPTEMBER

First return following the invention of the telescope. Observed between August and September. Halley saw the comet at this return, and the sighting prompted his calculations which discovered it to be identical with the comet of 1607 and 1531. He went on to predict its return in 1758.

1607 27 OCTOBER

Observed from September to October, nearly 5 weeks in all. Seen by Johannes Kepler. Very bright, reported as similar to Jupiter. Moved through Ursa Major, Boötes and Ophiuchus.

1531 26 AUGUST

Observed from August to September for a period of 5 weeks. Moved through Leo, Virgo and Libra. Very well seen. Notable observation by P. Apianus who showed that comet tails always point away from the Sun.

1456 8 JANUARY

Seen May to July, for 6 weeks. Very well observed. A plot by Paolo Toscanelli shows the comet moving through Ophiuchus, Corona Borealis, Boötes. Said to have a tail 60 degrees long or 120 times the apparent diameter of the full Moon. Seen during the time of the Turkish siege of Belgrade and supposedly 'excommunicated' by Pope Calixtus III as being an agent of the Devil, though evidence for this is rather suspect.

1378 10 NOVEMBER

Observed for 6 weeks during September and November. Chinese and European records exist. Not a favourable return.

1301 25 OCTOBER

Observed for 6 weeks from September to October. Seen by the painter Giotto di Bondone, who later used his impression of it as a model for the Star of Bethlehem on his fresco *The Adoration of the Magi*. (Incidentally, Giotto is the name given to the European Space Agence probe due to encounter the comet in March 1986.)

Well seen and brilliant. Its tail was said to be 70 degrees long, or 140 times the apparent diameter of the full Moon. The records for this return come from China and Iceland.

1222 28 SEPTEMBER

Through September and October. Thought to be an omen of the death of Philip Augustus of France. Records give a deep red colour to the comet.

1145 18 APRIL

Seen for a period of 10 weeks from April to July. Possibly depicted in an illustration which accompanies a twelfth-century illuminated manuscript, the Eadwine Psalter, a collection of the Psalms. The comet drawing was thought to have been added later.

1066 20 MARCH

Observed for over 9 weeks from April to January. Very bright. The head similar in luminosity to Venus. Thought to have been a portender of King Harold's downfall at the Battle of Hastings and shown on the Bayeux Tapestry.

AD 989 5 SEPTEMBER
Observed for 1 month in August and September according to Chinese records only.

AD 912 18 JULY
Observed for only 1 week in July. Very unfavourable. Some sources suggest that a possible fragment of the comet was observed by the Chinese earlier in the year, though this could well have been either another comet or a brilliant meteor.

AD 837 28 FEBRUARY
Most brilliant return on record. Observed for just over 4 weeks in March and April. Approached to within 6 million km (3.7 million miles) of Earth. Brighter than Venus, with a tail 95 degrees long; moved 60°W on April 9-10.

AD 760 20 MAY
Observed for about 1 month. Chinese records suggest that it was very brilliant.

AD 684 2 OCTOBER
Observed for 6 weeks during September and October. Records made in both China and Europe. A drawing of the comet appears in the Nuremberg Chronicles which were printed much later in 1493. The drawing is supposed to represent the comet at this return and is generally regarded as the earliest known depiction of Halley's Comet.

This appearance was connected with and blamed for a period of very bad weather and an outbreak of plague.

AD 607 15 MARCH
Observed for about 6 weeks from May to July. Chinese records very uncertain, with two comets observed around this time. Close approach to Earth on 19 April of only 13.5 million km (8.4 million miles).

AD 530 27 SEPTEMBER
Observed for 1 month in August and September. Chinese records again very poor. Reported as looking like a blazing lamp.

AD 451 28 JUNE
Observed for nearly 9 weeks from June to August. Chinese and European records made. Comet moved through Taurus, Leo and Virgo.

Shone down on the Battle of Châlons which brought defeat to Attila the Hun.

AD 374 16 FEBRUARY
Observed for about 9 weeks from March to May. A close approach to Earth occurred on 2 April when the comet was only 13.5 million km (8.4 million miles) away. According to Chinese records the comet was seen in Ophiuchus.

AD 295 20 APRIL
Seen in May. Chinese records very poor. Travelled through Andromeda to Virgo.

AD 218 17 MAY
Observed for about 4 weeks during April and May. Moved through Auriga, Gemini, Leo and Virgo, according to Chinese records.

Thought to be a portender of the death of the Roman emperor Macrinus.

AD 141 22 MARCH
Observed for just over 4 weeks in April and May. Close approach to Earth on 22 April when comet 25.5 million km (15.8 million miles) away. Reported bluish in colour.

AD 66 25 JANUARY
Observed for 10 weeks. Chinese and Jewish records exist. Seen as a sword-shaped star over Jerusalem and thought to be a portender of that city's fall shortly afterwards.

11 BC 10 OCTOBER
Observed for nearly 8 weeks from August to October. This is the first return for which we have reasonable records. Travelled through Gemini, Leo and Scorpius.

Return used for possible explanation of the Star of Bethlehem, but this now discounted.

86 BC 6 AUGUST
Records of this return very poor. Associated with trouble at the time of Octavius' consulship.

164 BC 12 NOVEMBER
Until 1984 there were no known records of this return, but in August of that year F. R. Stevenson, K. K. C. Yau and H. Hunger reported that they had found records of the return on some Babylonian tablets kept in the British Museum. Records of the 87 BC return were also found. It seems the comet was first seen in Taurus near the Pleiades and later in Western Sagittarius close to the planet Jupiter. Dates given are between September and November.

240 BC 25 MAY
Earliest reliable recording of Halley's Comet, regarded by many as the first true record. Mentioned by the Chinese.

466 BC MAY
Reports of two comets around this time in Chinese records from 467 to 465 BC. One may have been Halley but it is impossible to be certain.

1059 BC
Records very uncertain. All Chinese. Thought by some to be the earliest possible record.

Comet Halley may have been brighter in the past than it is today. Activity in the comet over the last few returns suggests that a great deal of material has been lost. It is therefore likely that the comet is suffering each time it approaches the Sun and may eventually fizzle out. It is just possible, however, that it may have some surprises in store for us before then.

THE LAST TWO RETURNS 1835 AND 1910

As we have seen, Halley has been recorded on at least 29 returns but most of these were not well documented. The last two returns have been the best observed, and both are worth covering in some detail.

THE 1835 RETURN

Following the observation of Halley in 1759, which was the first predicted return, that of 1835 was eagerly awaited. The date for the comet's perihelion passage had been worked out by two French astronomers, Baron Damoiseau and Comte de Pontécoulant, who gave 4 November and 13 November respectively. Pontécoulant turned out to be the closest: both had based their calculations on the 1759 return alone. A German astronomer, Otto Rosenberger, decided to try to compute a more accurate date using the return of 1682, and he came up with a date of 11 November, but despite his more intricate set of calculations he was still not as close to the correct date as Pontécoulant.

Predictions could never have been exact. The calculations took into account the gravitational effects of all the known planets, including the recently added Uranus, found in 1781. They could not, however, include the one remaining major planet, Neptune, which was not discovered until 1846, eleven years after the 1835 return of Halley. These outer planets can have a considerable effect on the motion of Halley if they happen to be in a part of their orbit close to the outer portion of Halley's orbit. On the other hand, if they are on the opposite side of the Sun their influence is not so great. Wherever they are their effect must be taken into account if we are to obtain precise calculations.

The English astronomer John Herschel, son of Sir William Herschel (famous for his discovery of Uranus), resolved to make a search for the comet from a temporary observatory which he had set up at Feldhausen in South Africa. He began his searches at the end of January 1835, but without success. On the night of 6 August 1835 two astronomers at the Collegio Romano Observatory, Francisco di Vico and Father Dumouchel, located the comet near the star Zeta Tauri. Perihelion actually occurred on 16 November, so the predictions though not entirely accurate were not too far out.

By 23 September the comet had become a naked-eye object with the beginnings of a tail. By the middle of October this tail had grown to 20 degrees in length and had become very conspicuous. During October details were seen in the tail which seemed to change rapidly from night to night. Obviously the comet was disturbed in some way.

As perihelion approached, the comet became increasingly difficult to observe. It was picked up after perihelion by Herschel in late December. Again the striking variations in the tail were evident, and in fact the comet seemed even more active following perihelion. Herschel's observations were the best and most detailed, and it was he who last saw the comet on 20 May 1836.

Despite the fact that Halley had been followed for a period of 9 months, very little was learned about it from this return. There had been quite a close approach by the comet to Earth on 10 October, when the distance between the two had been only 8 million km (5 million miles) roughly 20 times the distance of the Moon. The comet's motion was better understood, due to the long period for which it was held under observation, but spectroscopy was in its infancy and photography was a thing of the future. Actually John Herschel was one of the pioneers in photography, but its application to astronomy was a long way off and the first successful photograph of a comet would not be taken for another 47 years.

THE RETURN OF 1910

By 1910 photography was widely used in astronomy and it played a major part in the study of Halley's Comet at this return. More important, it was by photography that the comet was first located.

The date for perihelion had been predicted by two astronomers, A. C. D. Crommelin and P. H. Cowell, both of whom worked at Greenwich Observatory. In 1907 they gave their predicted course of events, with an accurate description of what could be expected from the comet. Their date for perihelion was only three days out.

First to locate the comet was Max Wolf at Heidelberg, in Germany. The photographic plate on which he found the image was taken on the night of 11 September 1909. Once located, images of the comet were found on photographs previously taken by other observatories, the earliest being one taken at Helwan Observatory in Egypt on the night of 24 August. From measurement of the comet's position on these plates, its accurate path through the sky could be worked out for the following months. The first plate taken from Helwan was loaned to Greenwich for measurements to be made. It was broken in the process, but fortunately the image of the comet was not damaged.

By the night of 16 September observations by E. E. Barnard of Yerkes Observatory, Wisconsin, gave a magnitude of 15.5. Over the next few nights the comet brightened slowly and fluctuations in its brightness were detected, which seems to be characteristic of Halley's Comet.

By the middle of November the comet was at 13th magnitude with an apparent diameter of 11 seconds of arc. This is quite small — about the same as the apparent diameter of Mars at an average opposition. By mid-December a tail about 2 minutes of arc long could be traced: that is about 1/15 the apparent diameter of the full Moon, so the comet remained quite small. Still most notable were the brightness fluctuations from night to night. By the end of 1909 Halley's Comet could be seen with small telescopes. But now it became overshadowed by the appearance of a much more spectacular non-periodic comet which turned up quite unexpectedly at the beginning of 1910.

The Great Daylight Comet, as it became known, was

Halley's Comet and Venus, 6 May 1910, taken with the 250mm (10 inch) Cooke lens at Johannesburg. (Courtesy Royal Astronomical Society)

discovered by a group of diamond miners in South Africa. It was the most brilliant of the twentieth century so far. By mid-January it could easily be seen with the Sun above the horizon. Its head was brighter than the planet Venus, and it had a tail which took the form of a great curve no less than 50 degrees in length or 100 times the apparent diameter of the Moon. It must have been an impressive sight, and it is probable that this comet was confused with Halley's by many people, since it was far more conspicuous and of course at the time the news was full of reports of Comet Halley. The Daylight Comet's period was so long that it is impossible to work out with any accuracy, though it is likely to be in the order of thousands of years. It had never before been recorded and will probably not be recognized as the same comet when and if it returns.

Halley eventually reached naked-eye visibility on 11 February and it was again Max Wolf, rediscoverer of the comet, who first reported this. During late February and early March the comet remained fairly inconspicuous because it was on the opposite side of the Sun from Earth. Remarkable activity still persisted in the comet, however.

Halley reappeared from out of the solar glare on about 12 April and rapidly became a conspicuous naked-eye object. Perihelion was reached on 20 April, and from then until early May the comet was a beautiful and striking object in the pre-dawn sky, making a fine sight with the brilliant planet Venus, also a morning object at that time. The magnitude of Comet Halley's head in mid-April was around +2.5 (about as bright as the pole star), and during the whole of April remarkable changes were taking place in the tail.

The motion of the comet brought it between the Earth and the Sun on 18 May, so once more for a brief period it was lost in twilight. The head of the comet was actually predicted to pass in transit across the Sun's disc on the eighteenth, and it was expected that if the nucleus were large enough (in the order of 100 km, 62 miles) it would be seen as a tiny black spot moving across the face of the Sun. No such thing was seen. At least this set an upper limit to the size of the nucleus; we now know that the nucleus is far smaller than this, but its true size is still not known.

One day later it was predicted that the Earth should pass through the comet's tail. It is doubtful whether this did actually happen but at the time the prediction caused considerable alarm, many people believing that they would be poisoned by gases from the comet. No ill-effects were experienced so far as is known, neither were there reports of any phenomena known to be associated with the event. Some people reported unusual colours in the sky and a few noticed an unusual smell on the night in question, but it is fairly certain that this had nothing to do with our encounter with cometary material.

Later calculations have suggested that owing to distortions within the comet tail and general lagging of dust particles, the tail probably missed us by 400,000 km (250,000 miles). This is actually a very small margin, and it is hard to see how such a precise statement can be made when dealing with something as uncertain as the true limits of a comet's tail. Some sources suggest that there was a delay and passage through the tail occurred on 20 May. The tail was certainly doing some interesting things at this time. On 17 May it had reached 75 degrees in length and at least 9 degrees in width, and on 18 May 75 degrees of tail could be seen sticking up from the horizon and it was estimated that possibly 30 degrees was hidden below the horizon. By 19 May the tail was no less than 140 degrees, extending almost across the whole sky.

The closest approach to Earth by Halley on this return was on 20 May, when it passed to within 21 million km (13 million miles). This was close but by no means a record. Though the spectacular views rapidly came to an end after this, the comet remained quite conspicuous through the latter part of May and tail lengths of 20-30 degrees were reported, 40-60 times the apparent diameter of the Moon.

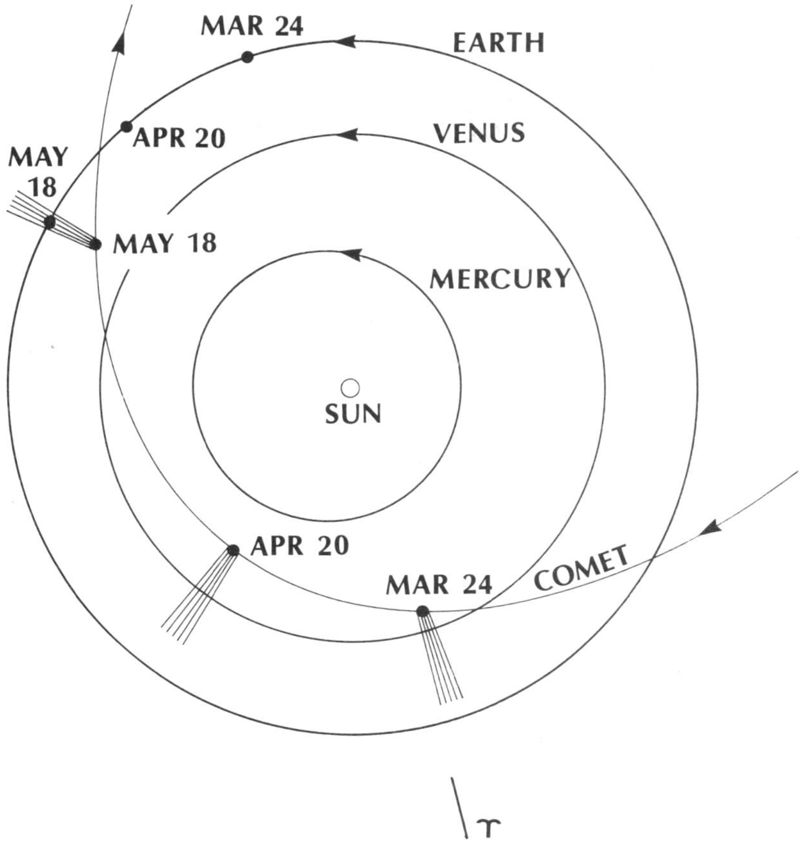

THE PASSAGE OF THE EARTH THROUGH THE TAIL OF HALLEY'S COMET IN 1910
In 1910 it was predicted that the Earth would pass through the tail of Halley's Comet. Though there is now some doubt as to whether this actually happened, the diagram shows the relative positions of the Earth and the comet during the 1910 appearance. It can be seen how this encounter with the tail might have occurred.
It is interesting to compare the circumstances of the 1910 apparition with those of the present (see figure on p.24).

On 6 and 7 June a break in the tail occurred. In fact what happened is that the old tail appeared to detach itself completely from the head, while a new one formed very rapidly. Disturbances in the tail continued through much of June, and as late as 22 June E.E. Barnard at Lick Observatory was still reporting changes in the tail even though the comet had faded to 8th magnitude and was well below naked-eye visibility.

By November 1910 Halley's Comet had faded to magnitude 11 and by January 1911 it was below 13th magnitude. Even then disturbances were still going on, the brightness fluctuating quite violently. The last visual sighting was made by Barnard at Yerkes on 23 May 1911 when the comet was about magnitude 16. However, photographs showed the image up to 30 May, the comet being then as far away from the Sun as Jupiter. It had been followed for 21 months.

Aphelion (furthest distance from the Sun) was reached in 1948, when the long journey back to the Sun for the 1985/86 return began.

Halley's Comet on 2 June 1910, taken with the 750mm (30 inch) reflector at Halwan. Note the interesting disturbed structure of the tail. The stars on the picture appear as streaks because the exposure time was 30 minutes; during that time the comet moved a little and in order to obtain a sharp picture the telescope had to follow it, causing the stars to trail. The length of those trails indicates the distance moved during the exposure time.
(Courtesy Royal Astronomical Society)

THE 1985/86 RETURN

What may we expect of the 1985/86 return of Comet Halley? Some astronomers have predicted that this will not be a particularly favourable return, but there will be things of interest happening and it will certainly be the most closely studied return so far. More telescopes will be turned on to it than ever before, and greater technology will of course be applied to it. With the development of spacecraft over the past 28 years, probes will bring better pictures than could ever be achieved by Earth-based study.

The story of the 1985/86 apparition really began on 16 October 1982 when the comet was located for the first time for this return. D. Jewitt and G. Danielson, at Palomar Observatory, California, used an electronic device known as the Charge Couple Device (CCD) at the prime focus of the 5 metre (200 inch) Hale telescope. Exposures of 8 minutes were given using the CCD. It is estimated that with the CCD objects can be detected which are 50 times fainter than is possible with ordinary photographic methods. This takes us below magnitude 25. The image of Comet Halley when first detected was magnitude 24.3, or an estimated 700 million times fainter than the pole star. Halley's position in the sky was 8 degrees northwest of the bright star Procyon in the constellation of Canis Minor.

At this 1982 rediscovery Halley had the distinction of being the faintest comet ever recorded, although even at the staggering distance of 1600 million km (990 million miles) it was not the most distant comet ever seen. This record is still held by Comet Stearns 1927 IV, which was last observed on 13 March 1931, almost 4 years after its perihelion passage and standing at a distance of 1700 million km (1055 million miles). It was however a record distance for an observation of Halley. I would be prepared to bet that after its perihelion Halley will be followed out to a greater distance than this.

The search for Halley's Comet began as long ago as 17 November 1977 at Palomar. Predictions of the comet's magnitude were then 26. Many other observatories joined the hunt, all hoping to be the first to locate it. Since recovery many observatories have been successful in picking it up as it has slowly brightened. Kitt Peak confirmed the recovery only four nights later, again using a CCD on the 4 metre (157 inch) Mayall telescope.

The first amateur astronomer to record Halley's image on a photographic plate this time round was Tsutomu Seki in Japan. He detected the 20.5 magnitude image with his 600 mm (24 inch) reflector on 22 September 1984. The first report of a visual sighting was made by amateur astronomer Stephen J. O'Meara, using the 600 mm reflector at Mauna Kea, in Hawaii. On 23 January 1985 he claimed to have observed the comet when at magnitude 19.6, and this was backed up by images obtained with a CCD on the 2.2 metre (88 inch) reflector of the same observatory, though carried out separately. O'Meara saw what he thought to be a faint misty patch passing a particular star and this tied in exactly with the CCD image. It seems that this observation was genuine and that he must be regarded as the first person to see the comet since E.E. Barnard's last observation of 23 May 1911, 74 years earlier. (O'Meara actually breathed pure oxygen during his observation, to enable him to retain his ability to see faint objects when observing at the high altitude of Mauna Kea.) At magnitude 19.6 Halley would have been around 1 million times fainter than the faintest star visible to the naked eye without a telescope. (Incidentally, I have every reason to believe O'Meara's claim, for he came to my rescue in 1980 when I made an observation that took astronomers by surprise. They were ready to jump on me until O'Meara made another observation which proved mine to be correct.)

The reason that the 1985/86 apparition is unlikely to be very favourable is the position of the comet with respect to the Earth and Sun at the time of perihelion. It will never

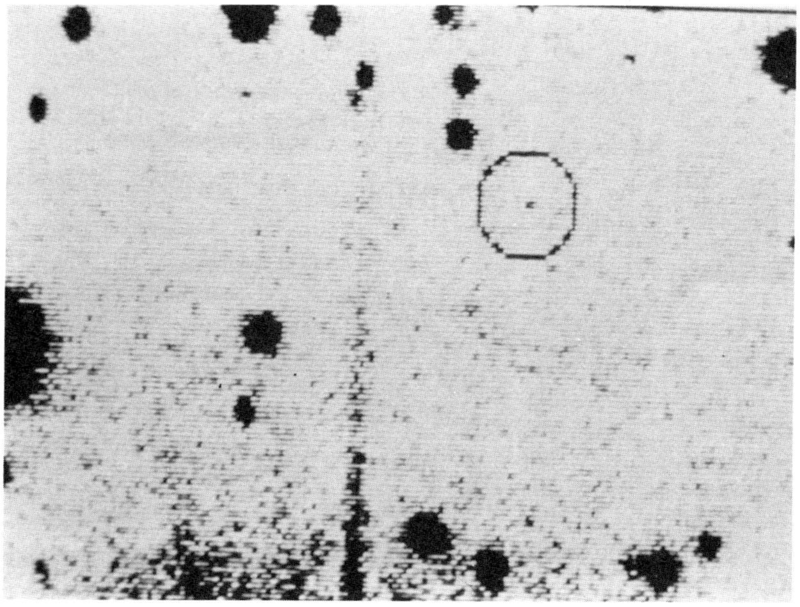

The recovery image of Halley's Comet obtained by astronomers using a CCD on the 5 metre (200 inch) Hale telescope at Palomar Observatory on 16 October 1982.
(Courtesy California Institute of Technology)

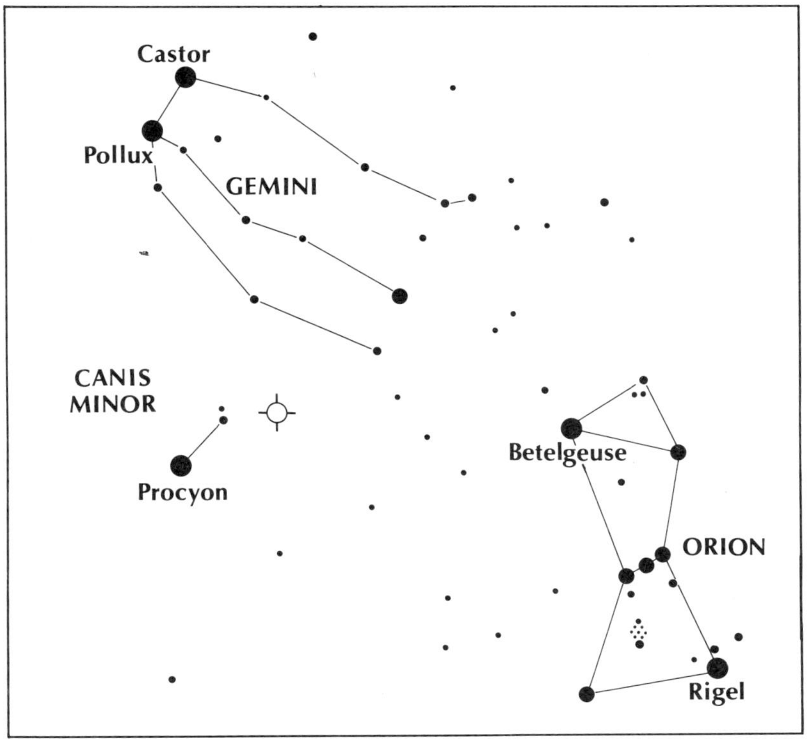

RECOVERY POSITION OF HALLEY'S COMET, 16 OCTOBER 1982
The actual position was Right Ascension 7h 11m 01.9s: Declination +9° 33′ 03″, which placed it in the constellation of Canis Minor near its border with Orion and Gemini, not far from the bright star Procyon.

actually come close as in 1910, and at perihelion it will be on the opposite side of the Sun to us.

There are some advantages, though. Because of this perihelion position we will see the comet well both on its way in towards and out from the Sun. A moderately close approach of 92 million km (57 million miles) will occur in November 1985, and in April 1986 it will pass 63 million km (39 million miles) from us.

As an interested observer, what you can expect to see is a brave show put on by a famous comet; you will not, unfortunately, see a brilliant head and long tail stretching right across the sky. However, you will have what previous generations never had, and that is the benefit of the barrage of technical and photograph equipment aimed at recording every possible aspect of this long-awaited event.

It is the instability of Halley, and in fact comets in general, that gives rise to two schools of thought on the expected brightness during the 1985/86 return. Optimists and pessimists disagree by as much as 2 magnitudes, which is a difference of over 6 times in brightness. One can never be certain just what a comet will do. As we have seen, in 1974 Comet Kohoutek, though predicted to be spectacular, proved to be a miserable failure. Comet West of 1976 was quite the opposite: though regarded as showing little promise it turned out to be one of the best comets this century. Therefore although we must accept that it is unlikely that Halley will be spectacular at this return, there is always the possibility that it could take us all by surprise.

Astronomers have had a long while to prepare their observational programmes, and they will not allow any opportunity to be missed.

The following is a calendar of expected events.

Between the beginning of 1985 and August amateurs will start to pick up the comet as a faint smudge of light, and this will gradually become easier to distinguish. Halley should be visible in moderate-sized telescopes of 300 mm (12 inch) aperture by the beginning of August 1985.

In the latter half of that month, with the absence of the Moon and resulting dark sky, it will be seen low in the east just before dawn, close to the 4th magnitude stars Chi 1 and Chi 2 Orionis, and not far from Zeta Tauri — near where it was first recovered in 1835 — known for its close proximity to the famous Crab Nebula (M1 in the Messier catalogue). The overall magnitude of M1 is 8.4 but its surface brightness is very low. It will be of interest to compare the comet with the Crab Nebula around this time, though the comet will be the fainter at magnitude 13.0. It will still be a difficult object even with moderate-sized telescopes.

SEPTEMBER 1985

With the disappearance of the Moon from the sky around mid-month, Halley will be found to have risen to magnitude 12. Though only 1 magnitude brighter than in August, this amounts to a total increase of 2½ times, so the comet should now be within range of most average amateur telescopes down to 250 mm (10 inch) aperture.

OCTOBER 1985

The magnitude of the comet will rise to 10 during the month, bringing it well within the range of 150 mm (6 inch) telescopes and probably smaller. During the month the comet will move through Taurus. After mid-month the moon will once more exit from this part of the sky, allowing for favourable viewing conditions. First signs of the comet's gas tail should show in larger telescopes and on photographs.

NOVEMBER 1985

The Moon will be absent throughout the middle of the month, which is fortunate because the comet will pass close to two beautiful star-fields during this period. Some very fine photographs should be possible.

On 11 November the comet will pass only 4°N of the Hyades star cluster, a grand V-shaped formation of stars forming the head of Taurus the Bull. Included in this field, though not physically a part of the Hyades, is the bright orange star Aldebaran, adding colour to the scene. At 7th magnitude the comet should be within the range of 10×50 or 7×50 binoculars, and since this is the best way to see the Hyades, the view should be most pleasing with this type of instrument.

Five nights later, on 16 November, Halley will pass just

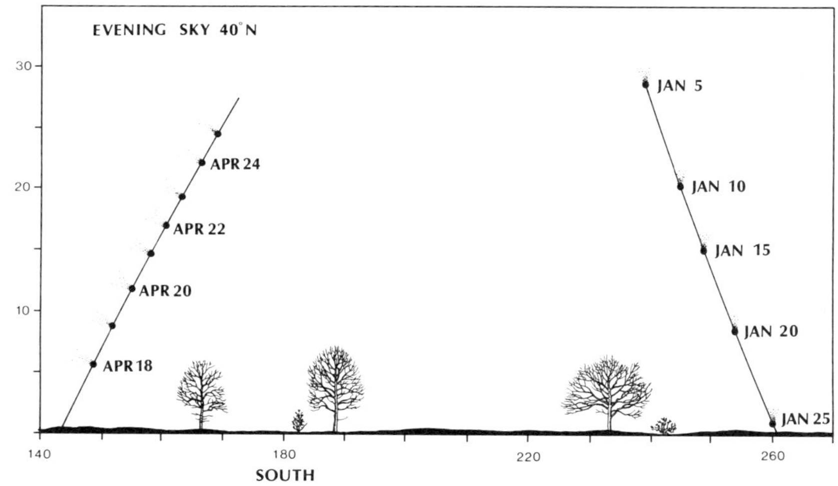

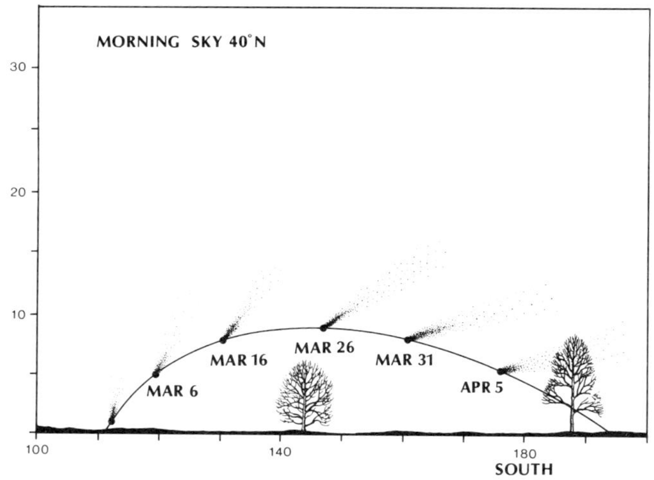

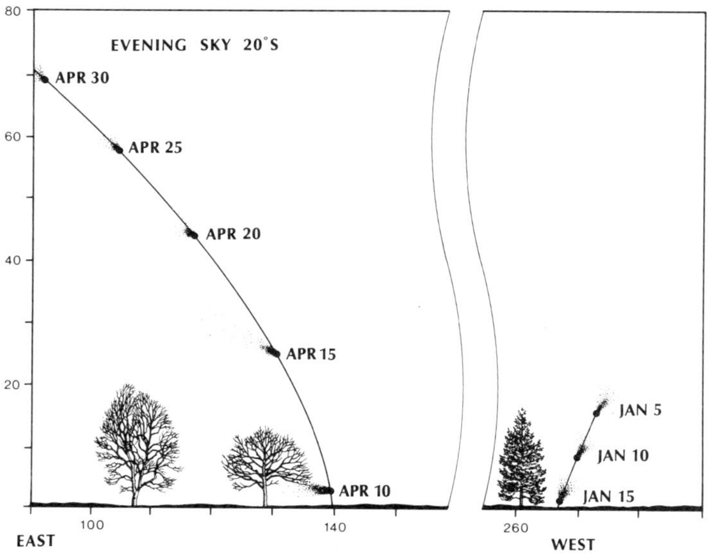

THE POSITION OF HALLEY'S COMET IN THE SKY DURING 1986
These four diagrams are for use at latitude 40° North and 20° South, but they can be adapted for latitudes either side of this by altering the position of the horizon. Charts for both morning and evening are given and are for periods when the comet is most prominent. Values along the bottom of the chart are for degrees in azimuth. Vertical values to the left-hand side are for degrees of elevation above the horizon.

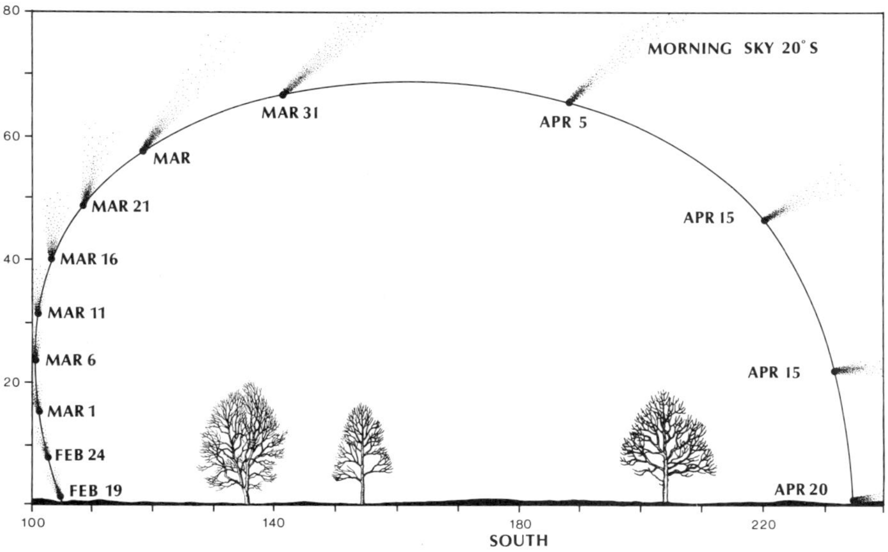

2°S of the magnificent Pleiades or Seven Sisters star cluster, again ideal for binoculars and making a charming and memorable sight.

On 27 November Halley will pass closest to Earth on its inward pre-perihelion journey. Now it will be found close to the 4th magnitude star Gamma Arietis. The comet is also close to opposition at this time (opposite the Sun in the sky), and the tail will therefore be pointing more or less away from us. It will also appear to swing from one side of the comet to the other as a result of alignment with Earth.

DECEMBER 1985

The first naked-eye sightings of Halley's Comet for 75 years are expected at the beginning of this month. The comet will then be in the constellation of Pisces, 50°W of the Sun.

The Moon will be troublesome at mid-month and it will affect the comet's visibility considerably. To see the comet easily with the naked eye it is essential to have a dark sky free of moonlight and to be situated well away from artificial lights. A haze-free sky is a must: even the slightest haze, especially if the comet is near the horizon, will make it very difficult, even impossible, to see.

After Christmas the comet will have attained 6th magnitude, still near the limit for naked-eye viewing. It will change in brightness very little now. The increase in the comet's actual brightness as it closes in to the Sun will be cancelled out by its increasing distance from Earth. During this last part of the month the comet will pass close to the star Gamma Aquarii. Binoculars should show approximately 5 degrees of tail, or 10 times the apparent diameter of the Moon. When viewed this way it should be quite a respectable comet.

JANUARY 1986

The comet will very soon disappear into evening twilight and this will be the last chance for observers in the northern hemisphere to see it well. On 13 January a

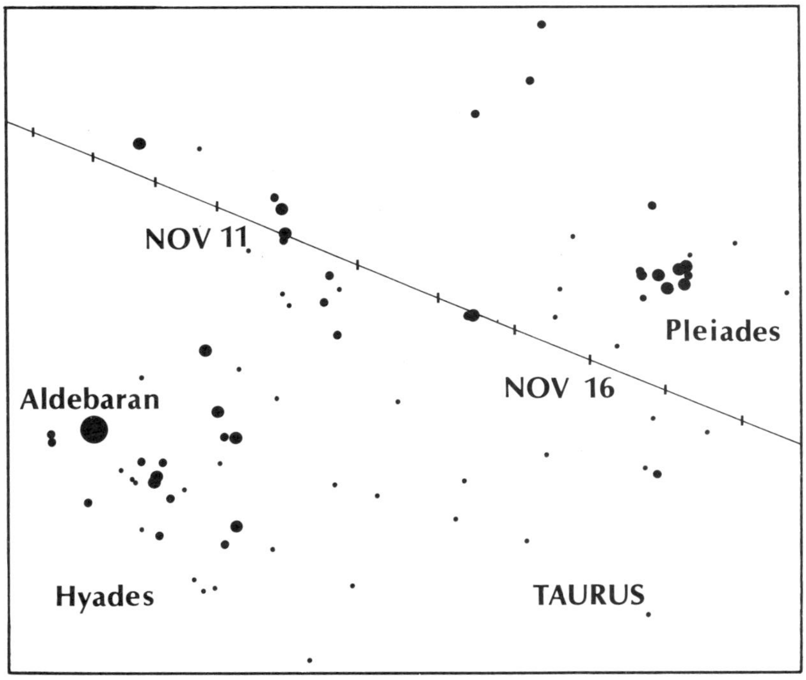

HALLEY'S COMET PASSING THROUGH THE HYADES AND THE PLEIADES IN NOVEMBER 1985
The dashes on the track indicate the position of the comet at Oh Universal Time each day, with the dates when the comet will be closest to each cluster. At this time the comet is expected to be around magnitude 7.5 so it should be visible with binoculars.

splendid spectacle may be seen as the comet passes close to Jupiter and the thin crescent Moon. Those in mid-latitudes should have the best view. The head will be of 5th magnitude, and first signs of the dust tail should be seen, making the tail generally much more conspicuous. It is not likely that any visual Earth-based observations will be possible after 24 January until the comet appears once more after perihelion.

FEBRUARY 1986

Comet Halley is due to reach perihelion on 9 February and only in the last few days of the month will Earth-based observers once again catch a glimpse of the comet — and then only those in the southern hemisphere. We will of course be able to follow the comet's progress through the various deep space probes and spacecraft in Earth orbit.

It is expected that first signs of the tail will be seen from Earth by 20 February. By 25 February the head itself should be seen and will be around magnitude 3, with a tail 10 degrees long. Pessimistic astronomers, however, feel that this is an over-estimate; they predict that it will not be more than magnitude 4.5. Halley has the habit of appearing much brighter after perihelion, so we must wait and see. Comets are, if nothing else, very unpredictable, and though they may sometimes disappoint us they can just as easily surprise.

MARCH 1986

Again the comet will rapidly approach Earth, and in the middle of the month it will be found in the constellation of Sagittarius with the Moon absent. It is hoped that the head will be around magnitude 3 and the tail at least 20 degrees long, possibly more.

On 13 March the European spacecraft Giotto makes its flyby of the comet, preceded by the two Russian Vegas on 6 and 9 March and the Japanese Planet A on 8 March. So though for many of us Halley will be inaccessible, it will be very much in the news and we may expect to see fine pictures returned from space.

APRIL 1986

It is during this month that the greatest expectations are held for the comet as seen from Earth. Optimistic astronomers expect the magnitude of the head to be 2.1, others predict only magnitude 4.0. The tail is expected to reach 40 degrees in length. It will be in the constellation of Centaurus by mid-month, and to observers in Australia and South Africa it will be overhead in a dark sky before dawn. This is certainly the best time to see the comet.

On 24 April a very unusual event will be seen by those in the Pacific area. The comet, now fading but still with a 10 degree tail, will be 40 degrees away from the full Moon which will go into total eclipse. For a while the sky will darken and the comet should show up clearly. At this time the Moon will be in Virgo close to the star Spica, and the comet will be found among the stars of Crater and Corvus.

MAY 1986

During May the comet will fade rapidly. There should still be signs of a tail, but by the end of the month it will fade to below naked-eye visibility. It will remain thus for the next 75 years.

With the help of moderate telescopic equipment it will of course be possible to follow the comet for a few months longer. Conjunction with the Sun will occur in August, with Halley around 9th magnitude. By the time the comet reappears from the Sun's glare only large telescopes will show it. This will be when most amateurs lose it, but I think it is safe to predict that professional astronomers may follow the comet until as late as 1990. If this is so, it will have been followed in total for a period of 7-8 years — which, when compared with a period of 21 months during the last return, shows the enormous technical advances made in astronomy over the last 75 years.

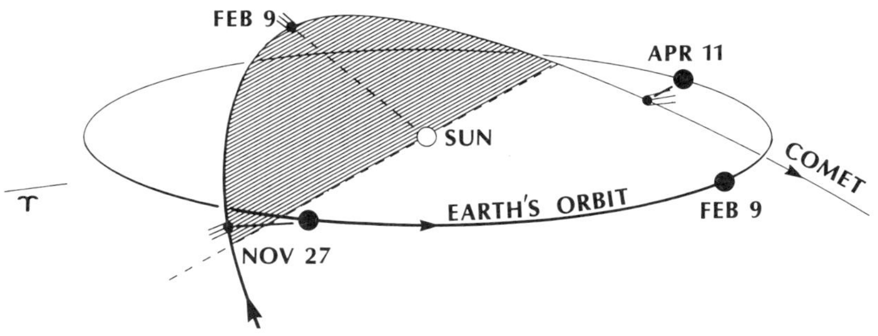

THE RELATIVE POSITIONS OF THE EARTH AND HALLEY'S COMET FOR THE 1985/86 RETURN

Note that at perihelion on 9 February 1986 the Earth and the comet will be on virtually opposite sides of the Sun. There will be, however, two fairly close approaches to the comet in November 1985 and April 1986.

SPACECRAFT TO HALLEY AND THE FUTURE

As already mentioned, the most exciting aspect of this return is that we are in a position to pay a visit to Halley, thanks to our recent entry into the space age.

Three nations are sending a total of five spacecraft to encounter the comet at close range, not only revealing much about Halley itself but teaching us more about comets in general than any other method so far devised. If all goes well we should see close-up pictures of the comet on our own TV sets in the first two weeks of March 1986, shortly after they are received from space.

The Russians are sending two of their Venera type spacecraft, called Vega (a Russian acronym of Venus, *Ve*, and Halley, *Ga*). The plan is for both craft to encounter Venus first. Then, using the alignment of that planet with the comet's path and the gravitational force of Venus, both probes will be flung on to encounter Comet Halley, making for greater economy in the mission. This principle is similar to that used by the American Mariner 10, Pioneer 11 and Voyagers 1 and 2 spacecraft, all of which visited more than one planet, using alignment and gravity to deflect them on to the next target with greater speed but using less fuel.

The Vega probes were successfully launched on 15 and 21 December 1984; they should encounter Venus in June 1985. Both probes will then release lander capsules and balloons into Venus atmosphere, sending information back to Earth from the planet's blisteringly hot surface. Meanwhile the two main spacecraft will bypass Venus at a distance of 30,000 km (20,000 miles), having made course corrections to set them on line for encounters with Halley.

On arrival at Halley, according to Russian sources, both spacecraft will release capsules to encounter the comet itself, passing at a distance of 10,000 km (6000 miles). Vega 1 will make its encounter on 6 March, Vega 2 on 9 March. If Vega 1 is destroyed by materials given off by the comet, Vega 2 can be programmed to pass further away, at 20,000 km (12,000 miles) distance. If Vega 1 is not destroyed, Vega 2 will pass within 5,000 km (3000 miles). Total flight times to Venus for each craft are calculated at 174 and 176 days respectively; if these are correct Vega 1 should arrive at the planet around 8 June, Vega 2 on 16 June 1985. The findings of the Vega craft will be used by scientists working on the Giotto mission in an attempt to make it more effective.

The Soviet probes will be loaded with instruments. These will include still and TV cameras, dust impact detectors, dust mass spectrometers, infra-red spectrometers, plasma analysers, energy particle detectors, plasma wave detectors, ion mass spectrometers, magnetometers and neutral mass spectrometers. This instrument package has been designed to study the physical chemistry or the comet's nucleus, head and tail. From this we should learn about the origin of comets, and also be able to establish if the nucleus is actually solid or just a condensed swarm of particles, as suggested by some astronomers. Image resolution should be in the order of 50 metres (165 feet).

Japan is also sending two spacecraft to study the comet. The first, called MS-T5 Sakigake or Pioneer, was launched

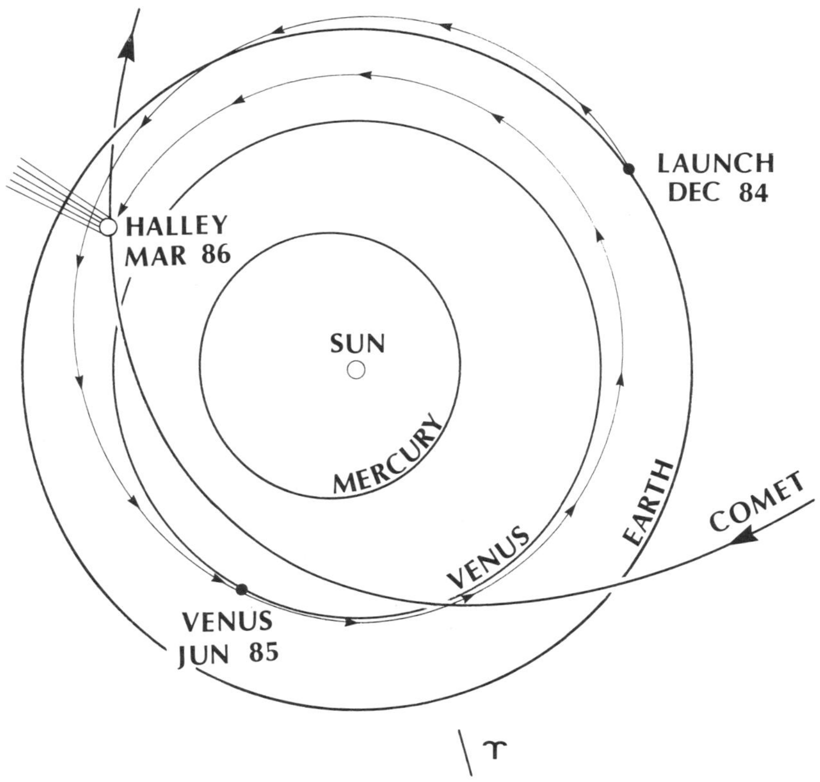

THE TRAJECTORY OF THE VEGA SPACECRAFT
The thin line shows the path of the two Vega spacecraft and how, following their encounter with the planet Venus, they will continue on for a further 9 months for an encounter with Halley's Comet on 6 and 9 March 1986.

1 HIGH GAIN ANTENNA
2 REACTION CONTROL THRUSTERS
3 LOW GAIN ANTENNA
4 MEDIUM GAIN ANTENNA
5 SOLAR ARRAY
6 ULTRA-VIOLET IMAGE CAMERA

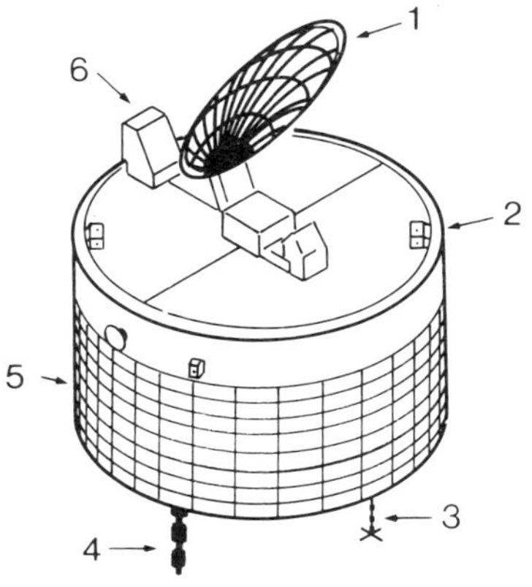

PLANET A

THE JAPANESE PLANET A SPACECRAFT
Planet A is a spin-stabilized craft 1.4 metres (4 ft 7 in) in diameter and 70 cm (27½ in) high. It is hoped that the closest approach to the comet will be in the order of 100,000 km (60,000 miles), though some sources believe it will be more like 2-3 million km (1.2-1.9 million miles).

The spin rate of the craft will be 120 rpm at launch, dropping to 6 rpm en route to encounter and 0.2 rpm while imaging on the comet.

Mitsubishi Electric made the bus structure and the space division of Nissan was responsible for the launch vehicle. The Institute of Space and Astronomical Science (ISAS) in Japan is responsible for co-ordination of the mission.

on 8 January 1985 and weighs only 138 kilograms (304 lb). This will not itself make a close encounter with the comet but it will be used to study solar wind as a back-up to the other Japanese probe, Planet A, which will approach the comet. MS-T5 is also testing the hardware to be used for Planet A.

Planet A is due to be launched on about 14 August 1985. Its approach to the comet will be the most distant of the close encounter missions at between 100,000 and 1 million km (60-600,000 miles) from the nucleus. This will take place on 8 March 1986 between the two dates of the Vega encounters. The combined comet and probe speed will be 70 km per second (43.5 mps). The main instruments on Planet A will be a magnetometer and an ultra-violet camera.

The last spacecraft to encounter the comet will be the European Giotto mission. It is to fly closer to the comet than any of the others.

Launch date for Giotto is set for July 1985 and will be aboard an Ariane launch vehicle fired from Kourou, French Guiana. The actual encounter will take place between 13 and 14 March 1986 and will last only 4 hours at a possible flyby distance of only 500 km (310 miles). At this time the comet will be 145 million km (90 million miles) from Earth and the flyby velocity will be 68 kps (42 mps). If the spacecraft encounters any cometary debris at this speed the result can only be catastrophic. It has been calculated that a particle weighing 1/10 gram would have the same kinetic energy as a 600 kg (1320 lb) object travelling at 100 km per hour (62 mph). The spacecraft will be so angled that any such impacts occur to one part of it only. This face is protected by a double shield which it is hoped will protect the spacecraft from impact by particles of this size.

Giotto will carry a camera which, because of the high flyby speed, will have to pan with the comet. Power is supplied by solar cells which cover the cylinder of the spacecraft body. It is expected that at its closest approach, pictures with a resolution limit of 20 metres (65 feet) will be obtained. Other instrumentation is similar to that on the Vega probes. At encounter the probe will weigh 512 kg (1126 lb) only 52 kg (114 lb) of which is actual payload. So, though far heavier than the Japanese probes, Giotto is less than half as heavy as the Soviet craft.

It is hoped that Giotto will show us the chemical make-up

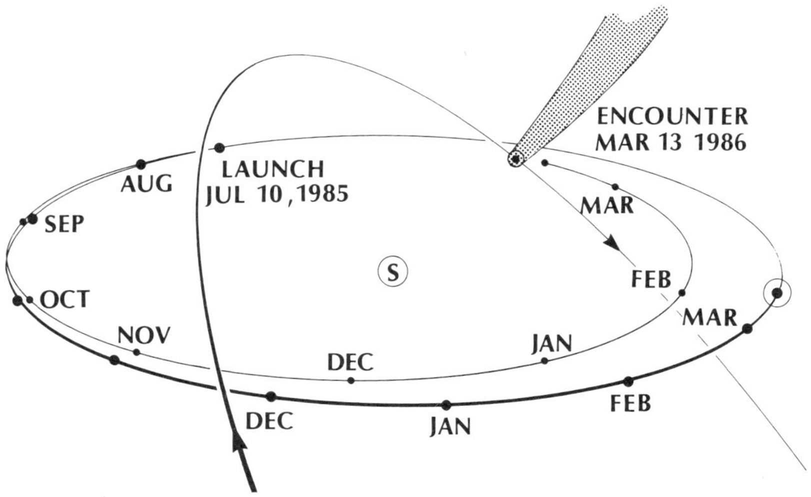

THE TRAJECTORY OF THE GIOTTO SPACECRAFT
The relative positions of the Earth and the spacecraft for the first day of each month and at encounter on 13 March 1986 (Earth circled), are shown.

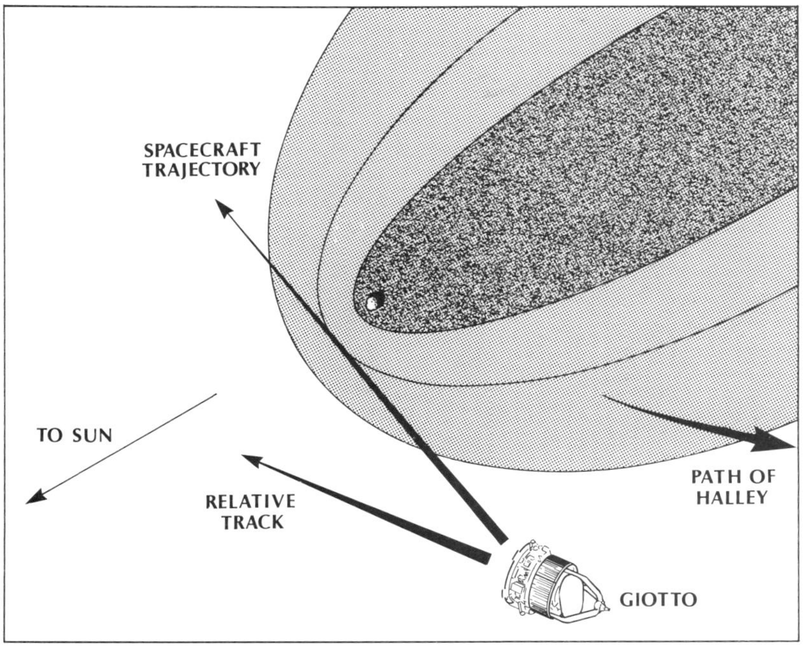

THE PATH OF GIOTTO THROUGH HALLEY'S COMET
Giotto is expected to pass within 500 km (300 miles) of the nucleus, with a combined comet-spacecraft speed of 68 kilometres per second (42 mps).

The Giotto spacecraft. Following initial system testing at British Aerospace Space and Communications Division the spacecraft was transported to the Centre Spatial de Toulouse for further tests.
(Courtesy British Aerospace)

1	X AND S BAND FEEDS
2	ANTENNA TRIPOD
3	DESPUN HIGH GAIN ANTENNA DISH
4	OPTICAL PROBE EXPERIMENT
5	CAMERA EXPERIMENTAL SENSOR
6	STAR MAPPER
7	NEUTRAL MASS SPECTROMETER SENSOR
8	BUMPER SHIELD TO PROTECT FROM IMPACTS
9	AXIAL THRUSTERS
10	SAFE/ARM UNIT
11	JOHNSTONE PLASMA ANALYSER
12	THERMAL CONTROL SHUTTER
13	ION MASS SPECTROMETER 1
14	ION MASS SPECTROMETER 2
15	EARTH-SUN SENSOR
16	PARTICLE IMPACT ANALYSER
17	RADIAL/TANGENTIAL THRUSTER
18	SOLAR PANEL
19	UPPER THERMAL SHIELD AND RADIATOR
20	MAGNETOMETER SENSOR
21	CARDIOID ANTENNA

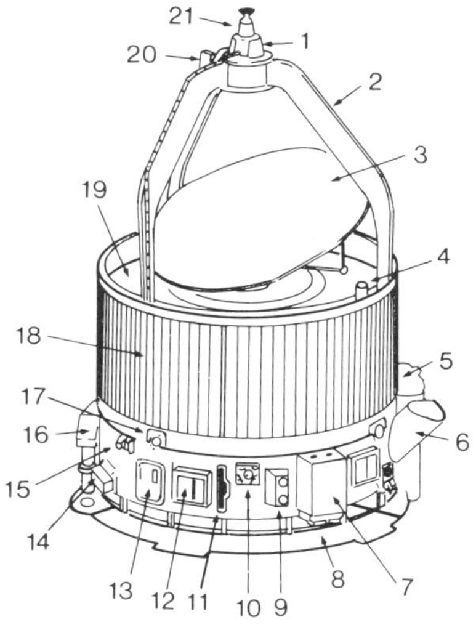

GIOTTO

THE GIOTTO SPACECRAFT
Giotto is a spin-stabilized spacecraft. The prime contractors for the spacecraft are British Aerospace (Bristol), and co-ordination of the mission will be carried out by the European Space Agency (ESA).
Dimensions of the craft are as follows:
Height: 3 metres (10 feet)
Diameter: 1.84 metres (6 feet)
Transmitter power: 20 watts
The 685 million km (425 million mile) journey will take 8 months.

of the coma and just what is going on within it. It should also tell us how much dust there is in the comet. Information on the flow of ionized gas and the effect on this of the solar wind will also be studied.

All the craft will be taking advantage of the period when the comet is expected to be at its most active — 130 million km (80 million miles) distant from the Sun, having passed through perihelion a month earlier.

Besides these special encounter missions there will be other spacecraft operating at the time that will be switched from their normal programme in order to study the comet. The United States, though it will not be sending an encounter mission, will not be out of the running by any means. Originally NASA had planned a very adventurous mission, taking in not just Halley but another comet called Tempel 2 in 1988. It was to be called the Halley Intercept Mission, and would have carried 125 kg (275 lb) of payload. Unfortunately the Government would not fund the mission. Despite an attempt to gain the funds through public support, the mission did not materialize; the money raised is being used to support US scientists involved in studies of the comet by means of other spacecraft.

NASA has several other projects planned. The Space Shuttle will be used for three missions beginning in March 1986. Eleven scientists have been selected to study the comet with Astro, a space-lab based astronomical observatory. Although the scientists must remain on Earth, three astronomers will fly with the Shuttle as Astro specialists. A detailed study will be made of Halley's tail and the nucleus. In addition to this the American probe Pioneer Venus, currently orbiting Venus, will also study the comet from a far better vantage point than here on Earth.

In spite of losing out on the Halley Intercept programme, America has still managed something of a scoop. A

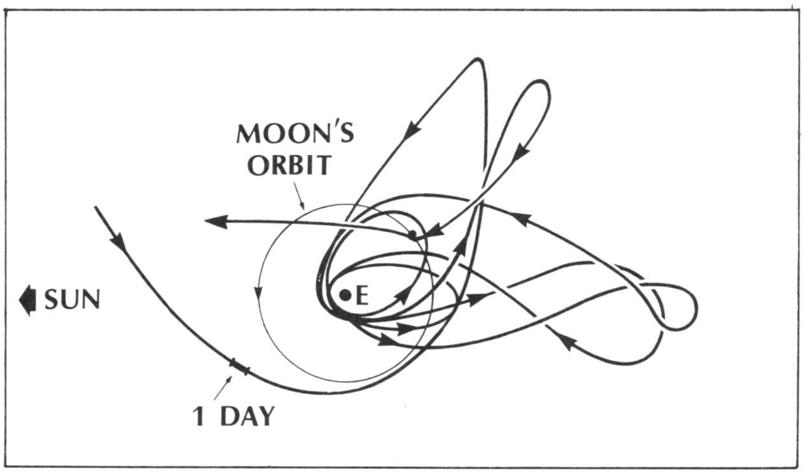

COMPLICATED MANOEUVRES OF THE INTERNATIONAL COMETARY EXPLORER (ICE)
The probe was commanded to begin this series of manoeuvres in June 1982. There followed a series of five lunar flybys beginning March 1983, and ending in December 1983 with a 116 km pass above the lunar surface. The probe is now heading for an encounter with Comet Giacobini-Zinner on 11 September 1985.

spacecraft, originally called ISEE-3, was sent up in 1978 to study charged particles from the Sun (solar wind). By a remarkable series of complicated manoeuvres, using the gravity of the Earth and the Moon and at one point approaching to within 100 km (62 miles) of the lunar surface, this has been sent out on a trajectory which will allow it to intercept a faint comet known as Giacobini-Zinner. The probe now goes under the name of ICE, or International Cometary Explorer.

The encounter with this comet will take place on 11 September 1985 and it will pass between 5000 and 15,000 km (3100 and 9300 miles) of the comet's head. In so doing it will be the first spacecraft to achieve such an encounter. But that is not all. The probe will also study Halley on two occasions, one on 31 October 1985 at a distance of 138 million km (86 million miles), and a second on 28 March 1986 from 31 million km (19 million miles). Reusing this spacecraft means a cost of only $5 million instead of the projected cost of $200 million for a special encounter mission. The probe will study gases and magnetic field, and it may also register the impacts of dust particles indicating dust density.

The next twelve months are certainly going to be significant for cometary science. When one considers the advances made in science since Halley's last visit it is difficult to imagine what position we will be in at its next return in 2061.

The obvious guess is that a manned mission to the comet will be made, provided that it is thought to be worth while and safe. Such a project would certainly be of interest and would attract great publicity. Scientifically more exciting but not so newsworthy would be an automatic mission to take samples of the comet material and bring it back to Earth. This would be a difficult mission, and the survival of Giotto will teach us much about its feasibility. It should be possible to obtain samples of material from the coma or tail, if not from the nucleus, and there would be no better way of learning of the comet's make-up and probable origin.

By Halley's next return we may however know all we need to know about comets; the event may be of nothing more than historical interest. No doubt some enterprising businessman will consider it profitable to send out a craft for samples in order to sell them. I am sure many people would pay a high price to own a piece of the most famous of celestial visitors.

RECORDING COMET HALLEY

Many people who have the opportunity to see Halley's Comet will wish to know how to record its appearance photographically. Many astronomers will already have built special equipment for this purpose. To record the comet well enough for the image to be of scientific value would require sophisticated equipment, a description of which is beyond the scope of this book. However, the layman can easily get a picture of the comet which though it may be of little if any scientific value, will certainly give immense satisfaction — particularly when shown to his grandchildren in preparation for the next visit!

In these days of fast lenses and very high speed film, no specialized equipment is necessary to record faint objects such as a comet. Provided that the camera can be guided in some way, to counter the movement of the stars due to the Earth's rotation, really good images of the comet will be obtained with exposures of about 5 minutes. With an F2 or F1.7 lens, using film ranging from ASA 400 to 1000, images of stars fainter than those detectable with the naked eye can be obtained with exposures of only 15 seconds. Fifteen seconds is the longest exposure time one can give with an unguided camera using a standard 50mm lens without the stars trailing due to rotation of the Earth. Under 15 seconds the stars will appear as dots, not as lines as they would with longer exposure times.

All that is needed is a tripod to hold the camera steady during the 15-second exposure time; a device for holding the shutter open for the required time; a fast film; and a dark clear sky away from city lights.

Good star pictures can be obtained even without the presence of a comet; if the photographs are made on slide film the colours of the stars will show up. If a bright comet happens to be around, delightful shots will result with very little effort. Pictures taken night by night will even show the comet's movement among the stars, and if near a particularly interesting star field, as Halley will be in the second week of November 1985, the photographs will be of great value to those who take them. If the comet displays a particularly long tail the photograph will show it more clearly than it can be seen with the naked eye.

Remember however that a really dark observing site is essential. Modern sodium street lighting is a curse to night sky photographers, and if you are not careful the sky in your picture will turn out orange. This will obscure any delicate detail in the comet's tail, thus losing the most spectacular part.

The more adventurous photographer can obtain good pictures by mounting the camera piggy-back style on to a telescope, provided that this has an equatorial mount. A clock drive is not necessary, though useful: keeping the object central in the eyepiece for 3 or 4 minutes by hand guiding will result in very good images. Stars down to magnitude 8 or 9 can be captured in this way, and tails of comets well beyond their visual limits. A very useful aid is an eyepiece with a crosswire, for guiding on the comet. If your telescope has a motor drive synchronized with the Earth's rotation, it will be easier still, but even so a close watch must be kept to ensure that the guiding does not go astray.

For most of us this will be our only chance to see Halley's Comet, so a little bit of effort is well worth while.

Author's own photograph of Comet Iras-Araki-Alcock, taken on 11 May 1983. The photo shows what interesting results can be obtained with the minimum of equipment. An ordinary SLR camera was used with a 135mm telephoto lens. The print was made by combining two negatives taken 5 minutes apart. The image of the comet is lined up but the star images are doubled. This is due to the comet having moved during the time between the two exposures, the distance between the double images of the stars showing how much movement had occurred. The four bright stars to the left of the comet form part of the Plough, or Big Dipper (Ursa Major).

Many non-astronomers ask how they may best see the comet and what sort of telescope they should get. Unfortunately, unless one is prepared to lay out considerable amounts of money, most of what can be obtained in the way of a telescope is of little use and will certainly lead to disappointment. The alternative is to build one yourself, though this of course takes time and a degree of skill.

Without doubt the best form of investment for the casual observer is a pair of 10×50 or 7×50 binoculars, that is, binoculars with a 50mm aperture with 10 or 7 times magnification. If the comet is very bright with a very long tail, binoculars are probably the most practical type of instrument due to their wide field of view and light grasp. They are also quite cheap to buy and are useful for other subjects besides astronomy.

Some of the most successful comet hunters use binoculars. George Alcock, a British amateur astronomer famous for his discovery of several comets and nova, uses a pair of 25×105 binoculars. Large binoculars such as these do need a tripod, however, because they are very heavy. Their usefulness in a true scientific sense is illustrated by Mr Alcock's success.

METEORS AND THEIR RELATIONSHIP TO COMETS

Observation of comets can be made other than through direct study of the comet itself. Here no telescope is necessary and, provided the observations made are systematic and part of a planned observing programme, they can also be of scientific value.

Comets, as they move round the Sun, can be very seriously affected and lose quite a lot of material through the boiling off of volatiles. Emitted along with the volatiles are masses of dust particles. These are ejected into space and continue to move around the Sun in orbits similar to that of the parent comet.

If the orbit of those particles meets the Earth we are treated to a shower of meteors. As Earth moves through the debris, dust particles burn up on entering the atmosphere. The average rate for non-shower meteors, or sporadic meteors as they are called, is 5 per hour. When a shower is in progress the rate of meteors per hour increases, and the difference between the shower rate and the sporadic rate is the true shower rate.

There are many of these meteor showers through the year, but only a few are really active — notably the Perseids in August and the Geminids in December. The names are usually derived from the constellation which is the centre of the meteor shower radiant.

As the Earth travels through space and encounters the dust particles, they seem to radiate from a certain part of the sky into which the Earth is heading. So not only does the increase in number of meteors signify that a shower is in progress, but when the path of the meteors is traced back many can be seen to originate from a particular region of the sky.

Many of the streams have now been associated with certain comets. The first person to establish this in 1872 was the astronomer G. Schiaparelli, famous for his observations of the Martian canals. The best known of all the meteor streams is the Perseid shower which has been connected with Swift-Tuttle, a comet with a predicted period of 120 years. In 1980 there was a sudden increase

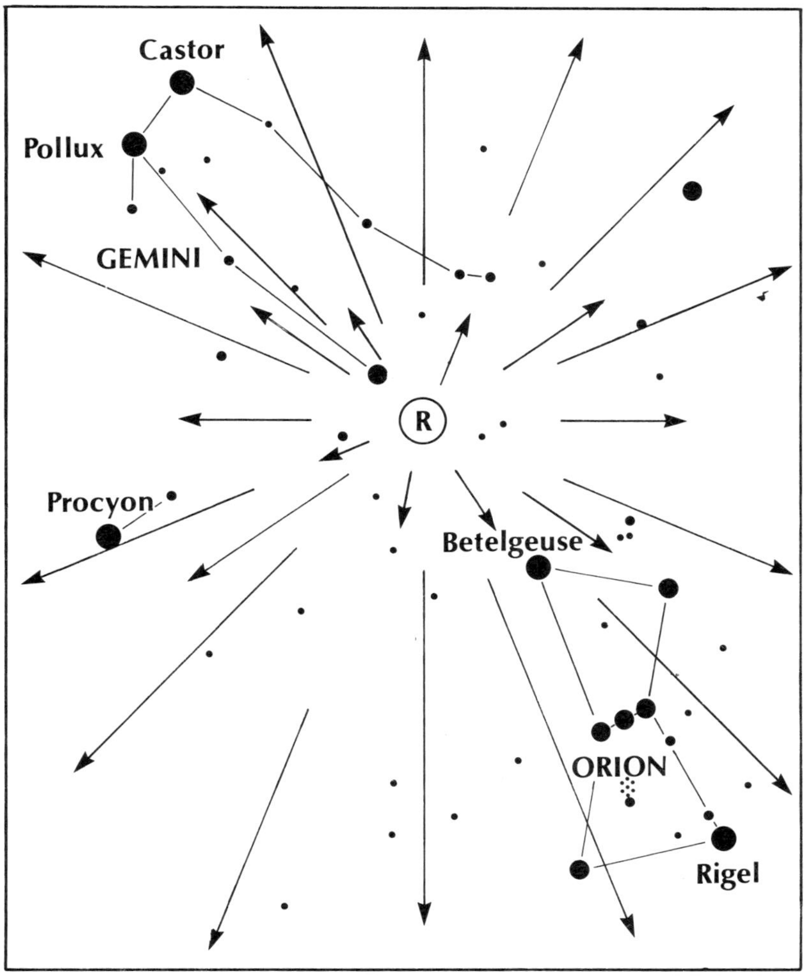

METEOR RADIANT IN ORION
The circle marked R centres on the general region from which the Orionid meteors appear to radiate.
The number of meteors that may be expected at maximum activity is about 20. There is a double maximum with this shower, the dates of which are around 19 and 27 October, though some shower meteors can be expected between 12 October and 4 November.
The other meteor stream associated with Halley's Comet, the Eta Aquarids, occurs in the southern hemisphere around 24 April to 20 May, with peak activity on 5 May.
The diagram gives fictional meteor tracks showing how one might expect a plot of tracks to look after 2-3 hours of observation.

of activity in the shower, and this was thought to be coincidental with the return of the parent comet predicted for 1982. Perseid rates were watched closely for the next two years, but no further increase in activity was noted. Furthermore the parent comet was not recovered. The conclusion was that either the predictions were in error and the comet came earlier than expected, or the comet has broken up and the 1980 increase in meteor activity was a sign of this.

Halley's Comet has its own associated showers. In fact there are two — the October Orionids and the April-May Eta Aquarids, the latter a southern hemisphere shower. Both are really the same stream, with the Earth encountering the orbit of the debris in two different places. Obviously any increase in the activity of these two showers is eagerly awaited. Concentration of debris is likely close to the comet itself. It is therefore expected that when the Earth passes through the orbit at a time when the comet is nearby or has recently passed, meteor frequency will increase.

An extreme example of this concentration is the Leonid shower which occurs around 17 November and is associated with Tempel's Comet. This has a period of 33 years: the meteor shower, which is usually poor, suddenly bursts into activity every 33 years to coincide with the return of the comet. Rates of 100,000 meteors per hour have been recorded in the showers of 1799, 1833 and 1866. The years 1899 and 1933 proved a disappointment, but there was another spectacular shower in 1966. Unfortunately the densest part of the shower seems only to be visible in very localized regions.

The comet Giacobini-Zinner produced a major shower in October 1946 when the Earth passed through the comet's orbit only 15 days after the comet itself. Another comet called Biela was seen to split in 1846. At its next return in 1852 the two components were further apart. After this return they were never seen again, but in 1872, a year when the comet should have returned once more, a brilliant meteor shower was seen. Evidence points to the fact that the comet had disintegrated. Another search for the comet is due to be made during 1985 — a last-ditch effort to see if the comet still exists.

Interest in the two showers associated with Halley is obviously considerable now and will continue to be so over the next few years. Any increase or lack of increase will be of scientific interest.

No increase in activity is expected. It appears that the Halley meteor streams result from debris left behind by the comet in ancient history. Since the orbit of the comet has changed since then, what we are seeing is dust left behind by the comet very many years ago. If this debris was deposited by the comet long enough ago, it will have spread itself around the whole of its old orbit quite evenly with no concentration coinciding with the comet's presence. This is the theory; if there is an increase in activity the whole subject will have to be looked at again.

There seem to be multiple sources of meteor activity in the case of this stream, and a better understanding of the position of these sources could tell us a great deal about the past orbital movements of Halley. Simple observation could therefore prove of great value. Meteor showers are easy to observe and no special equipment is needed. All one has to do is watch the sky in selected regions and log the meteors, giving details of their flight direction and brightness, time (Universal Time) and possible radiant. From a great number of such observations the sources of the stream may be accurately determined, affording us a better understanding of Halley's past.

Thus valuable observations of Halley can be made indirectly, with no instrumentation — just patience and a clear sky. So why not have a go? You won't have the chance again.

METEOR STREAMS AND ASSOCIATED COMETS

21 APRIL	April Lyrids	Thatcher 1861 I
5 MAY	Eta Aquarids	P/Halley
2 AUGUST	Alpha Capricornids	P/Honda-Mrkós-Pajdusákouá
11/12 AUGUST	Perseids	P/Swift-Tuttle
10 OCTOBER	October Draconids	P/Giacobini-Zinner
20/21 OCTOBER	Orionids	P/Halley
4 NOVEMBER	Taurids	P/Encke
17 NOVEMBER	Leonids	P/Tempel-Tuttle
21/22 DECEMBER	Ursids	P/Tuttle

There are two other active showers which are associated with no known comet. These are the January Quadrantids and the December Geminids.